LANDSCAPES
of UTAH'S
GEOLOGIC PAST

A Summary of Utah's Fascinating Geologic History

Thomas H. Morris, Kinsey G. Spiel, Preston S. Cook, & Hannah M. Bonner

A GEOLOGY UNFOLDED, LLC Publication

GEOLOGY UNFOLDED - Landscapes of Utah's Geologic Past: prepared by Thomas H. Morris (Ph. D.), Kinsey G. Spiel, Preston S. Cook, and Hannah M. Bonner; published by BYU Press under Copyright to GEOLOGY UNFOLDED, LLC 2016. Reorders: email at geologyunfolded@gmail.com or see us on Facebook at www.facebook.com/geologyunfolded. Photo of Earth as shown on the cover and this page from www.everystockphoto.com. See creativecommons.org.

ISBN: 978-0-8425-2997-6

Table of Contents

Preface

The first satellite images of Earth from space filled viewers with wonder. Never before had humans been able to see their planet from this perspective. Much was to be learned from analyzing the scale and shape of geomorphic features and interpreting the processes that created them. Fifty years later, we can track human-made and natural changes in landscape by viewing a set of images of a given area arranged in chronological order. This half-century view reveals that Earth is a dynamic planet. What would we learn if we were able to arrange satellite images of Utah in chronological order back to the Age of Dinosaurs? What landforms, mountains, lakes, and oceans have come and gone in that time? How did these events contribute to today's landscapes? In this book we attempt to address these questions. But instead of processing film from satellite-borne cameras we process geological data from the rock and fossil record to reconstruct past landscapes with amazing clarity. We can use these geological snapshots to interpret processes that shaped Utah in its geologic past and that have conspired through time to create today's spectacular scenery.

In order to complete these snapshots we have relied on the meticulous work of hundreds of geoscientists throughout the past century. By understanding the processes that formed correlative rock units, we can interpret ancient climates, topography, and major structural events. Supervolcanos, orogenies, interior seaways, and dune-filled deserts are some of the puzzle pieces of Utah's past. When we carefully put the puzzle pieces back together in time and space, a picture emerges. This picture is a recreation of Utah's ancient landscapes. This book recreates Utah's landscapes from seven distinct intervals of time starting in the Pennsylvanian Period and ending in the recent past ice age. These seven intervals are divided into chapters.

To make Utah's geologic history more visual we studied satellite imagery of modern Earth from around the globe and patterned our illustrations in a like manner. Occasionally artistic license was employed to illustrate numerous geologic features that occurred during slightly different time increments but still within a particular geologic period. These paleogeographic maps are the heart of this book and provide great insight into the past.

Because plate tectonics and the hydrologic cycle are the primary drivers of where rocks form, their deformation history, how they erode, and what landscapes they create, we have placed Utah into its plate tectonic and climatic settings for specific intervals of geologic time. These settings are summarized in the first paragraphs of each chapter. The type of depositional system that deposited sediment to form strata was dictated by the climate and geography. In the *Examples* section of each chapter we elucidate how tectonics affected specific areas and formations around Utah, especially in its national parks. Each chapter also includes the following:

- "Fast Facts" – discusses age, climate, key localities, common fossils, and provides a global map of the world's continental configuration at the time of interest,
- interesting life forms that existed,
- stratigraphic columns of named formations and rock types,
- outcrop maps illustrating where rocks of a given period are exposed in Utah today,
- "Did you know..." statements of interest,
- photos from Utah that exhibit key rocks from each interval of time, and
- bolded words referenced in the Glossary.

We hope the reader finds this book a succinct, engaging, and informative summary of Utah's geologic history. We also hope that it inspires greater curiosity into Earth's history and a deeper appreciation of this great and rare planet that we call home.

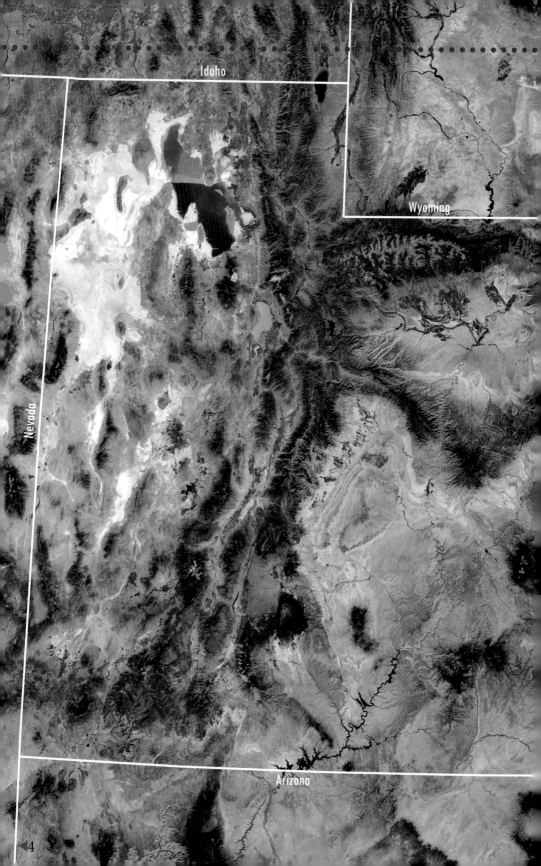

Idaho

Wyoming

Nevada

Arizona

Introduction

▶ If we were to look down upon Utah from a satellite orbiting Earth, we would observe a very colorful landscape: red colors in the south, white in the northwest, green in highland areas, and patches of beiges and blacks in the west. Why? What makes these landscapes? What past events were involved in creating these landscapes? Why are Utah's landscapes so awe-inspiring to so many? These are the fundamental questions that this book will explore. But first we must understand the basics of any landscape.

*Current satellite image of Utah from LANDSAT. Note variations in color and texture representing topographic, lithologic (rock type), and structural variation. Utah must have changed dynamically to produce these variations.

Landscape Evolution

▶ Landscapes result from the **hydrologic cycle** working on exposed rocks at Earth's surface. We must start with the understanding that ultimately most of the topography of the continents and ocean basins is created by Earth's crustal plates moving and interacting with one another. This process is called **plate tectonics**. The rigid outer crust of Earth is comprised of approximately a dozen large pieces or plates and many smaller ones. These plates are dynamic. Their motion is due to convecting heat deep within Earth's interior layers. As these crustal plates move around, they interact with one another forming volcanoes, mountain belts, and basins: essentially Earth's topography.

The next step in developing Utah's landscapes involves creating thick successions of sedimentary and igneous layers (strata). When rocks from the crustal plates are exposed to ocean or atmosphere, **weathering** breaks down the rock allowing gravity to **erode** materials from highlands to areas of low relief. This is greatly aided by the hydrologic cycle. Newly formed sediment accumulates through a variety of depositional systems. Deposition is centered in topographic low areas where the sediment may become buried and again turned to stone. Thus, differential relief and prolonged basin subsidence create thick accumulations of sedimentary strata, which can then be penetrated by igneous rocks. Sedimentary rocks comprise approximately 75% of the rock exposed on Earth's surface and most of Utah's surface area.

Once sedimentary rocks are created, they are usually deformed by continued tectonic forces. Under compressional forces they may fold and **thrust** over themselves. Under extensional forces they may break apart by **normal faulting**.

Finally, the deformed sedimentary and igneous rocks are differentially weathered and eroded. Harder rocks resist these processes and tend to stand out in relief relative to softer rocks (Fig. 1.1).

In summary, any landscape results primarily by:

1. creation of relief on Earth's surface by tectonic forces and the hydrologic cycle,
2. accumulation of sedimentary strata through a variety of depositional systems (**deltas**, rivers, sandy deserts, etc.) and potential igneous activity,
3. deformation of that strata by tectonic forces, and
4. **uplift** and differential weathering and erosion.

Climate may further overprint the landscape. In rainy and humid regions, vegetation grows ubiquitously, covering the rocks and hiding their colors. In arid and semi-arid regions, as in Utah, vegetative cover is reduced or nonexistent, allowing the red, green, gray, purple, black, and white colors of a wide variety of rock types to appear.

Fortunately Utah has many attributes that help create inspiring landscapes: a long history of plate tectonic activity that helped create thick successions of both sedimentary and igneous rocks of different age, multiple depositional systems creating a variety of rock types, active tectonics and erosion that provides present-day relief, and arid to semi-arid climates that produce little vegetation. If Utah can be viewed in light of these attributes, one can more fully appreciate its rare and wonderful landscapes!

Figure 1.1 *Aerial photograph of Comb Ridge in southeast Utah. Comb Ridge is a* **monocline** *on the east flank of the Monument Upwarp. Note the changing dip of the strata in the foreground and the* **laccolithic** *igneous intrusives of the Abajo Mountains in the background (skyline). This landscape illustrates many of the attributes discussed in the text including deposition of sedimentary strata, deformation, igneous intrusion, uplift, and differential weathering and erosion. View to the north. Photo courtesy of Ken Hamblin.*

Utah's Earliest History

▶ As we go further back into geologic time the rock record becomes increasingly scarce and therefore interpretation of geologic history is more tentative. This book focuses on Utah's geologic history from the Pennsylvanian Period to the present because the rock record is more complete and therefore better understood (Fig. 1.2). However, there were three major geologic events that pre-dated the Pennsylvanian that are worthy of mention. Utah's oldest rocks (~2,500 million years ago [Ma]) are found in the Raft River Range and Grouse Creek Mountains in northern Utah. This Precambrian rock is composed of metamorphic schist and igneous granite that represent accretion of the early North American continent. At approximately 750 Ma another major event in Utah occurred. The ancient supercontinent called Rodinia split apart and in Utah created a rift basin wherein a thick succession of sedimentary sandstones and mudstones were deposited. These rocks became hardened through time and presently are exposed in the elevated Uinta Mountains (Fig. 1.3). During the early Paleozoic (Cambrian through Mississippian - 541 to 323 Ma) Utah was mostly on the shallow marine shelf of the western edge of the North American continent. There it accumulated mostly shallow marine limestone, dolomite, and sandstone. Thus, these three events represent the underpinnings of the rock record that we examine in more detail.

Figures 1.4 and 1.5 should help the reader understand stratigraphic columns and the present-day geologic provinces of Utah.

Figure 1.3 *The strata comprising the heart of the Uinta Mountains are hard sandstones and mudstones of the Uinta Mountain Group. Sand and mud filled a rift basin some ~750 Ma which was eventually deeply buried, lithified, and then uplifted, re-exposed, and glaciated at Earth's surface.*
Photo of Mount Powell courtesy of Ken Hamblin.

Geologic Time Scale

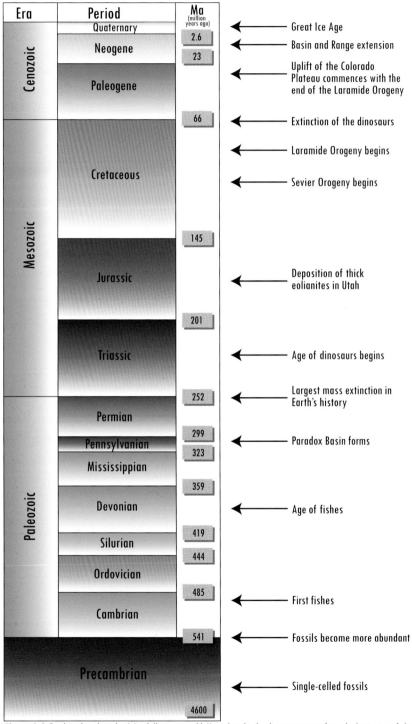

Era	Period	Ma (million years ago)	
Cenozoic	Quaternary		← Great Ice Age
	Neogene	2.6	← Basin and Range extension
	Paleogene	23	← Uplift of the Colorado Plateau commences with the end of the Laramide Orogeny
		66	← Extinction of the dinosaurs
Mesozoic	Cretaceous		← Laramide Orogeny begins
			← Sevier Orogeny begins
	Jurassic	145	← Deposition of thick eolianites in Utah
	Triassic	201	← Age of dinosaurs begins
		252	← Largest mass extinction in Earth's history
Paleozoic	Permian		
	Pennsylvanian	299	← Paradox Basin forms
	Mississippian	323	
	Devonian	359	← Age of fishes
	Silurian	419	
	Ordovician	444	
	Cambrian	485	← First fishes
	Precambrian	541	← Fossils become more abundant
			← Single-celled fossils
		4600	

Figure 1.2 *Earth is thought to be 4.6+ billion years old. Note that this book covers events from the beginning of the Pennsylvanian Period (~323 Ma) to the present.*

Guide to Understanding Stratigraphic Columns

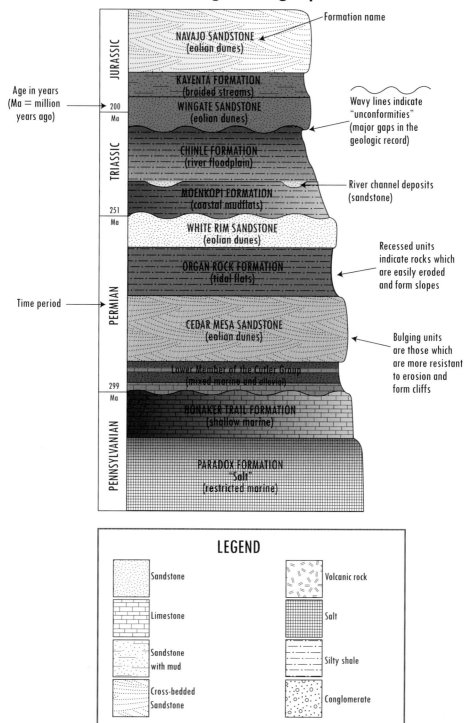

Figure 1.4 *Stratigraphic columns provide the reader with names of local formations, rock type, age, color, relief, environment, and other aspects of the local strata.*

Geologic Provinces of Utah

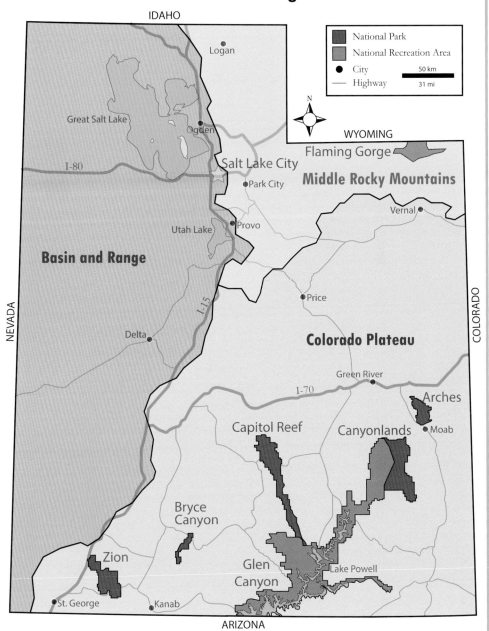

Figure 1.5 *Utah contains portions of three of North America's many geologic provinces including the Colorado Plateau, Basin and Range (Great Basin), and the Middle Rocky Mountains.*

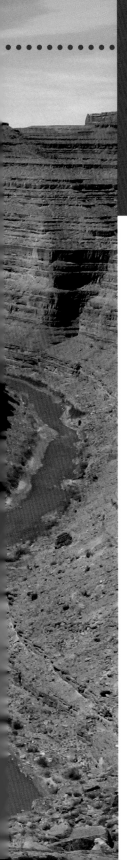

Pennsylvanian & Permian
323-252 MA
Uplifts & Basins

▶ During Pennsylvanian (323-299 Ma) and Permian (299-252 Ma) time the world's continents were assembled into a giant supercontinent called **Pangea** (see world map, pg. 17). The southern portion of Pangea was positioned over the south polar region which allowed large accumulations of continental-scale ice sheets to develop during cool conditions. As climate fluctuated, these ice sheets waxed and waned by building up on the land and then melting back into the ocean. As a result, sea level fluctuated dramatically. Elsewhere, large areas of Pangea were positioned near the equator providing climatic conditions conducive to making **peat** swamps and eventually coal deposits. Much of the world's coal was created during the Pennsylvanian Period. During the Pennsylvanian, Utah experienced arid climates and dramatic crustal deformation. During **lowstands** of sea level, eolian dune fields covered much of Utah. Tectonic uplifts extended from Utah to Oklahoma and have been called the Ancestral Rockies although they have little connection to the modern Rocky Mountains. Areas of uplift eroded while other areas subsided and filled with sediment. Often these two conditions are "yoked" wherein subsidence occurs directly adjacent to uplifts. Yoked uplifts and subsiding basins result from crustal changes associated with broad plate tectonic movements. Tectonic folding and thrusting often thicken the crust regionally. The thrust sheets acts as a regional weight which causes the crust to subside, like a bowling ball on a mattress. Basin subsidence is especially pronounced along the fringes of the uplift. Often regional drainage and marine waters inundate the low areas, filling them with water and effectively creating a mote around the edges of the uplift. When restricted, the water becomes very saline, depositing massive amounts of evaporites (salt and anhydrite). The uplift itself erodes through time, shedding its sediment into the adjacent sedimentary basin. Thus, it is the sedimentary strata in the basin that documents the past existence of a major uplift or mountain belt.

Cyclic strata of Pennsylvanian age along the San Juan River near Goosenecks State Park, southeastern Utah. Photo courtesy of Scott Ritter.

Pennsylvanian/Permian - Paleogeographic Reconstruction

Paradox Formation (~310 Ma)

A: Oquirrh Basin
B: Weber Shelf
C: Emery Uplift
D: Uncompahgre Uplift
E: Paradox Basin
F: Algal reefs on southwest shelf

Fast Facts

Age: Pennsylvanian Period 323-299 Ma
Permian Period 299-252 Ma
Climate: Repeated global glacial cycles, arid climate in Utah
Key localities: Goosenecks State Park, Arches and Canyonlands National Parks, Monument Upwarp, Wasatch Mountains
Common Fossils: Crinoids, brachiopods, phylloid algal mounds/reefs, lycopods, seed ferns

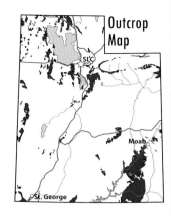

Outcrop Map

Stratigraphic Column - *Monument Upwarp*

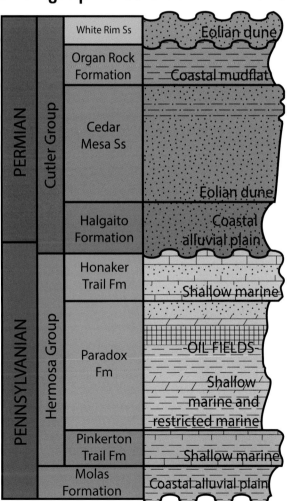

		White Rim Ss	Eolian dune
PERMIAN	Cutler Group	Organ Rock Formation	Coastal mudflat
		Cedar Mesa Ss	Eolian dune
		Halgaito Formation	Coastal alluvial plain
PENNSYLVANIAN	Hermosa Group	Honaker Trail Fm	Shallow marine
		Paradox Fm	OIL FIELDS / Shallow marine and restricted marine
		Pinkerton Trail Fm	Shallow marine
		Molas Formation	Coastal alluvial plain

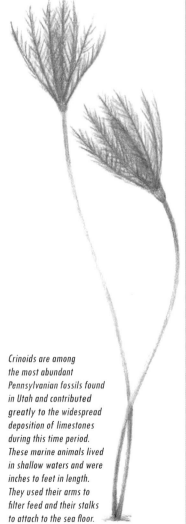

Crinoids are among the most abundant Pennsylvanian fossils found in Utah and contributed greatly to the widespread deposition of limestones during this time period. These marine animals lived in shallow waters and were inches to feet in length. They used their arms to filter feed and their stalks to attach to the sea floor.

Examples

▶ An example of a yoked uplift and basin is the Uncompahgre Uplift which was yoked to the Paradox Basin (see Paleogeographic Reconstruction, pg. 14). The Uncompahgre resulted from Utah's crust getting "jammed" by the advance of the Marathon/Ouachita **orogenic** belt from the southeast and island arc convergence from the west. The Ouachita's resulted from the collision of the ancient continents of North America, Europe, and Siberia with the supercontinent **Gondwana** from the south. As the Uncompahgre Uplift grew, silica-rich sediments in the form of **arkoses** accumulated along the basin's edge nearest the uplift. Eventually the Paradox Basin was flooded by marine waters from the south. As marine waters rose and fell through time due to glacial fluctuations in southern Pangea, thick accumulations of halite, gypsum, and **potash** salts accumulated in the arid Paradox Basin (Fig. 2.1). At the same time **carbonate rocks** and algal reefs accumulated on the basin's edge to the southwest. Algal reefs of the southwest shelf eventually trapped petroleum in a number of places including Utah's largest oil field, the Greater Aneth Oil Field.

Figure 2.1 *Potash evaporation ponds near Dead Horse Point State Park. Solution mining of subsurface Pennsylvanian salt is evaporated at the surface to produce potash used in fertilizer. A blue dye is added to the brine to increase evaporation. Photo courtesy of Ken Hamblin.*

Salt accumulations formed during the Pennsylvanian were deformed ductilely (like squeezing a tube of toothpaste) by accumulating Permian and Mesozoic sediments. The salt essentially flowed into large **salt walls** and **salt domes** (**diapir**; Fig. 2.2). Later, these salt structures and associated folded strata were buried by flat-lying Paleogene and Neogene sedimentary rocks. Finally, over the last six million years, the Colorado River established itself on the Colorado Plateau creating a drainage pattern indicative of the younger flat-lying sedimentary rocks. Once established, that pattern became **entrenched** and the river continued to erode the entire sedimentary succession of Neogene, Paleogene, and Mesozoic rocks. The river system then removed literally thousands of feet of strata from the Colorado Plateau in conveyor belt-like fashion. With time, the river **downcut** to the depth of the salt structures and overprinted or **superimposed** its established drainage pattern across the structural trend of the salt walls and domes. Thus, rather than follow the bends and folds of the salt structures, the river cut across these structures creating a visual "paradox" to early explorers (i.e., Paradox Basin). Tributaries to the Colorado River have now carved into the folds of the salt structures which are well exposed in the Moab Valley and Arches National Park area.

Figure 2.2 *The distinct white gypsum caprock exposed at the surface of the Onion Creek salt anticline near Moab extends over 6 square kilometers and attests to the subsurface movement of salt in the area. Photo courtesy of Ken Hamblin.*

Pennsylvanian World Map
★= Utah

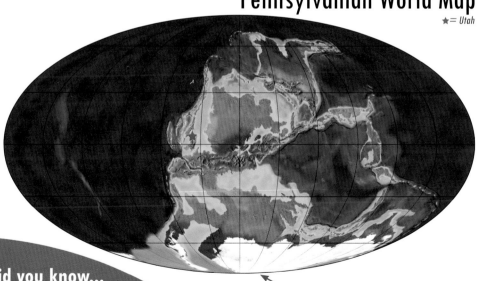

Supercontinent Pangea

Did you know...
the Pennsylvanian makes up less than 2% of geologic time, but most of the coal on Earth was deposited during this period.

Figure 2.3 *Southeastward view of the Raplee Anticline. The Raplee Anticline is a monoclinal flexure (see inset below) associated with the Monument Upwarp during the early stages of the much younger Laramide Orogeny. It exposes Pennsylvanian through Permian strata. Photo and inset courtesy of Ken Hamblin.*

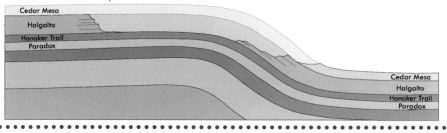

In north-central Utah the Oquirrh Basin subsided on what was essentially the **continental shelf** of western North American. Carbonate and **siliciclastic** sediment accumulated during subsidence. In some places the resulting strata is estimated to be more than 5 kilometers thick. Those strata are now exposed on Mount Timpanogos and Cascade Mountain in Utah County. During relative lowstands of sea level, much of the eastward (and landward) shelf was sub-aerially exposed allowing **ergs** to develop periodically. Today those ergs are recorded in the rocks of the Weber and Tensleep Sandstones of Utah and Wyoming. This ancient continental shelf in Utah is called the Weber shelf (see Paleogeographic Reconstruction, pg. 14).

Arid conditions remained for much of the Permian. In eastern Utah land surfaces continued to accumulate wind-blown sands of the Weber and Cedar Mesa Sandstones (Fig. 2.3). Westward marine waters inundated Utah to deposit the Toroweap Formation. At maximum flooding during latest Permian time, the Kaibab Limestone, which caps the southern rim of the Grand Canyon, developed across much of Utah. Upwelling marine waters in northeastern Utah created the Park City and Phosphoria Formations. Phosphate from these formations is mined for fertilizer.

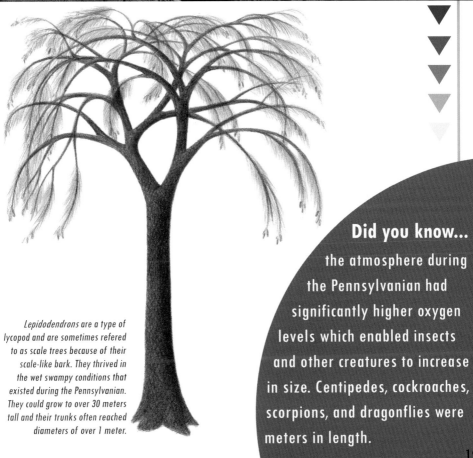

Lepidodendrons are a type of lycopod and are sometimes refered to as scale trees because of their scale-like bark. They thrived in the wet swampy conditions that existed during the Pennsylvanian. They could grow to over 30 meters tall and their trunks often reached diameters of over 1 meter.

Did you know...

the atmosphere during the Pennsylvanian had significantly higher oxygen levels which enabled insects and other creatures to increase in size. Centipedes, cockroaches, scorpions, and dragonflies were meters in length.

Triassic
252-201 MA

Extinctions & Breakup of Pangea

▶ The Triassic Period (252-201 Ma) is the first period of the Mesozoic **Era** (Fig. 1.2). The ancient continents of the northern hemisphere and the supercontinent Gondwana closed during Pennsylvanian and Permian time creating the supercontinent Pangea. Pangea was surrounded by the **Panthalassic Ocean**. Pangea existed into Triassic time but then started a slow breakup: first into the supercontinents Gondwana and **Laurasia**, and eventually into our present continental configuration. At the end of the Permian, a **mass extinction** of marine life occurred as much of the world ocean became toxic and uninhabitable for the majority of marine species. **Anoxic** conditions resulted from a complex, cascading series of events that included a lack of low-latitude mountainous uplifts (**orogeny**) and associated reduction in nutrient supply to the oceans, increased atmospheric CO_2 and global warming, melting of continental ice and resulting global marine transgression, etc. As a result, marine fossil species in rocks worldwide change dramatically across the Permian/Triassic boundary. Geoscientists recognized this change and used this major extinction event to divide the Paleozoic (ancient life) from the Mesozoic (middle life) Era.

*Early Triassic rocks of the Moenkopi Formation (bottom red beds), Late Triassic rocks of the Chinle Formation (middle), and Late Triassic/Early Jurassic rocks of the Wingate Sandstone (top cliff-former) along the Scenic Drive at Capitol Reef National Park.

Triassic - Paleogeographic Reconstruction

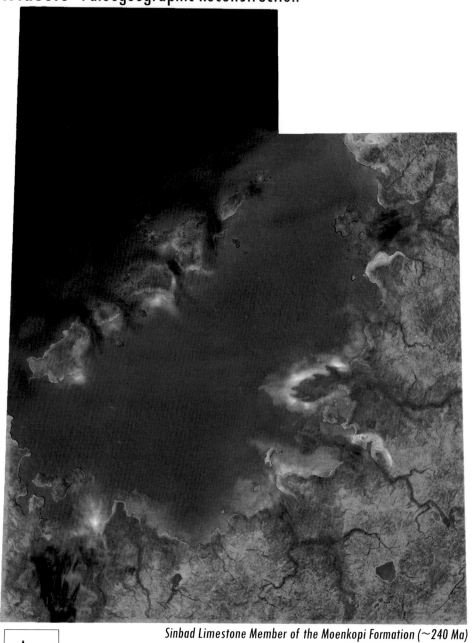

Sinbad Limestone Member of the Moenkopi Formation (~240 Ma)

A Panthalassic Ocean (deep marine)
B Carbonate banks
C Restricted shallow marine
D Muddy coastal plain and tidal flats

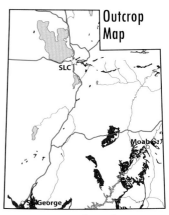
Fast Facts

Age: Triassic Period 252-201 Ma
Climate: Humid subtropic to arid
Key localities: Capitol Reef and Canyonlands National Parks, San Rafael Swell, St. George, Monument Upwarp
Common Fossils: Marine organisms such as mollusks, brachiopds, ammonoids, and fish. First lizards and dinosaurs began to appear.

During the Triassic Period, Utah was situated near the equator and on the western edge of the North American continent. It was pulling northward moving through humid tropic and subtropic climatic belts and eventually into the dry **trade-winds belt**. The western continental edge at this time was considered a **passive margin** with no dramatic plate tectonic activity and associated orogensis. The open Panthalassic Ocean was located to the west. The rocks of the Moenkopi and Chinle Formations reflect the hot and occasionally humid climatic conditions for most of the period. The Petrified Forest Member of the Chinle Formation exemplifies the **alluvial** lakes and swamps that existed in Utah through much of the Triassic when it was more humid. However, by latest Triassic time dry conditions of the trade-winds belt permeated Utah. Presently these trade-winds belts are located between 10 and 30 degrees north and south of the equator. Therefore Utah had drifted significantly north of the equator by Late Triassic time in order to have produced the vast **eolian** dune system (erg) of the Wingate Sandstone (Fig. 3.1).

Figure 3.1 *The Castle is a geomorphic feature created within the fractured Wingate Sandstone. The Wingate spans the Triassic-Jurassic boundary. The Castle is located directly across Highway 24 from the Capitol Reef National Park Visitor Center.*

Examples

▶ Rocks of the Early Triassic Moenkopi Formation represent the interplay between relatively flat alluvial and coastal mudflats with the open shallow marine waters of the Panthalassic Ocean. Numerous reptilian trackways (Fig. 3.2) carved and preserved in the bottom of shallow river channels attest to reptilian animal life in these wet alluvial plains (Fig. 3.3). As sea level rose and fell, marine carbonates were created during **highstands** of sea level, evaporates were deposited during sea-level fall, and red mudstones were created during the lowstands of sea level. Repeated fluctuations of sea level created cycles of these three interbedded deposits (**facies**). This is illustrated well in southwestern Utah where three carbonate-rich members (Timpoweap, Virgin, and Shnabkaib) are interbedded with mudstone-rich members (Lower Red, Middle Red, and Upper Red Members) of the Moenkopi Formation. In central Utah, on the San Rafael Swell, the Sinbad Limestone records the most eastward incursion of the Panthalassic Ocean into Utah (see Paleogeographic Reconstruction, pg. 22). This would be the last time central Utah would see the open ocean to the west. Some beds of the Sinbad Limestone are comprised of **oolitic** rock, which in modern oceans are created in warm, shallow, and clear marine waters such as on the Bahama Banks.

Extensive marine environments in the Triassic made for an abundance of ammonoids in Utah, and are commonly used as an excellent index fossil. They lived in open water and were good swimmers.

The Late Triassic Chinle Formation sits **unconformably** above the Moenkopi Formation as few rocks of Middle Triassic age are preserved throughout most of Utah. Alluvial plains with lakes and swamps were indicative of the non-marine terrestrial conditions that dominated Late Triassic time. Within the Petrified Forest Member of the Chinle, petrified trees two feet in diameter and tens of feet long are fossilized remnants of the warm, humid conditions that existed. Occasionally sediments eroded from southern uplifts would be transported northward by **braided river** systems to create the conglomeratic sandstones of the Shinarump Conglomerate and Mossback Members of southern and central Utah, respectively.

By latest Triassic time Utah moved into the trade-winds belt which was conducive to hot, dry conditions. The wind-blown quartz sand dunes of the Wingate Sandstone started to inundate Utah from the north. The Triassic-Jurassic boundary has been defined as being in the lower part of the Wingate. This period boundary is based on the discovery of Triassic dinosaurian trackways (**ichnofossils**) found in the lower part of the Wingate Sandstone. The dry, sandy, desert conditions of the latest Triassic Period continued throughout much of the Jurassic Period (Fig. 3.4).

Figure 3.2 *Reptilian tracks in the Torrey Member of the Early Triassic Moenkopi Formation at Capitol Reef National Park. Note the orientation of the individual prints is at an angle to the trackway path. This indicates the animal being carried by the current as it attempted to cross the river.*

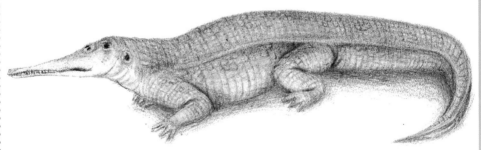

Figure 3.3 *Rutiodons were creatures similar to modern crocodiles. These carnivores reached lengths of 3 to 8 meters. They lived in swamps and rivers and would ambush prey that approached the water.*

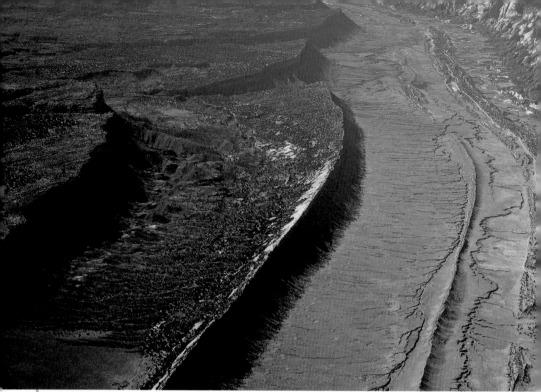

Figure 3.4 *Southward aerial view of the Strike Valley, Circle Cliffs Uplift portion of the Waterpocket Fold in Capitol Reef National Park. The orange rocks exposed on the right represent ancient deserts of wind-blown Triassic/Jurassic Wingate Sandstone, whereas the white band represents eolian sand dunes of the Jurassic Navajo Sandstone. Photo courtesy of Ken Hamblin.*

Stratigraphic Column - *Capitol Reef National Park*

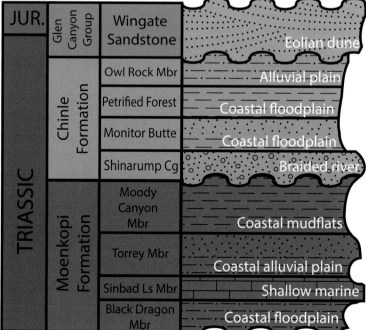

JUR.	Glen Canyon Group	Wingate Sandstone		Eolian dune
TRIASSIC	Chinle Formation	Owl Rock Mbr		Alluvial plain
		Petrified Forest		Coastal floodplain
		Monitor Butte		Coastal floodplain
		Shinarump Cg		Braided river
	Moenkopi Formation	Moody Canyon Mbr		Coastal mudflats
		Torrey Mbr		Coastal alluvial plain
		Sinbad Ls Mbr		Shallow marine
		Black Dragon Mbr		Coastal floodplain

Triassic World Map

★ = *Utah*

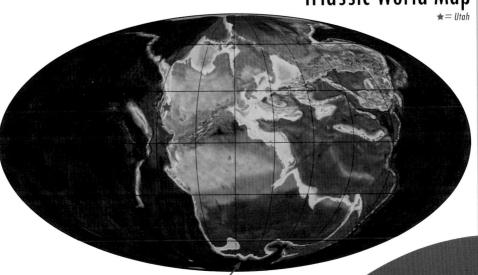

Supercontinent Pangea

Jurassic
201-145 MA

Sandy Deserts, Shallow Seas, & Dinosaurs

▶ North America and Utah continued moving northward during the Jurassic Period (201-145 Ma). Utah was firmly engulfed within the hot, dry trade-winds belt situated between 10 and 30 degrees north latitude. As a result, ergs dominated the landscape. Deposition of sandy deserts created the world famous Wingate and Navajo Sandstones and by Middle Jurassic time, the Entrada Sandstone. The North American Plate also started to move westward in a relative sense. This westward movement created plate **convergence** between western North America's continental crust and parts of the Pacific oceanic crust. This convergence resulted in a long north-south oriented **volcanic arc** and highland known as the Cordilleran Arc. This arc weighted the western margin of the North American continental crust and initiated long-lived tectonic features. These tectonic features included the Cordilleran highlands to the west and a "yoked" retroarc foreland basin to the east. The foreland basin included a foredeep and an occasional **ephemeral** forebulge and backbulge basin (Fig. 4.1). The foredeep was occasionally filled with marine waters. This yoked system of western mountains and eastern basins continued to migrate eastward through the Jurassic because of continued plate convergence. By Late Cretaceous time, this system culminated in an elevated mountainous thrust and fold belt in central Utah and an associated interior seaway in eastern Utah.

*Photo of Checkerboard Mesa in Zion National Park. The Jurassic Navajo Sandstone is composed of trough cross-stratified dunes that have been fractured by superficial weathering processes. Photo courtesy of Ken Hamblin.

Jurassic - Paleogeographic Reconstruction

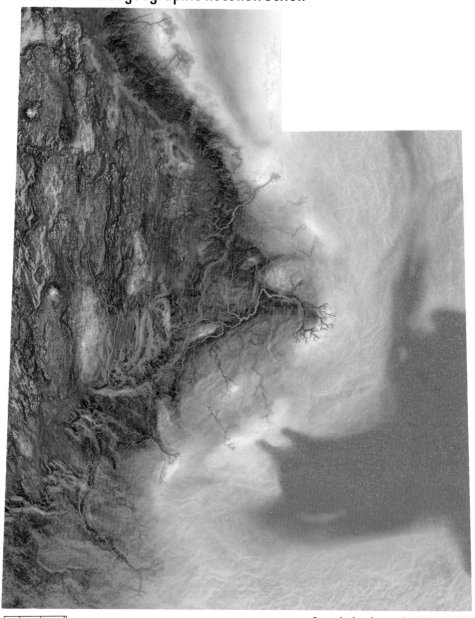

Entrada Sandstone (~165-161 Ma)

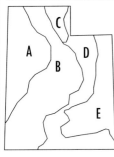

A Cordilleran Arc/Elko Orogeny
B Alluvial Plain
C Sundance Sea
D Sabkha/Inland Sabkha
E Wind-blown dunes (erg)

Fast Facts

Age: Jurassic 201-145 Ma
Climate: Arid
Key localities: Utah's Colorado Plateau and parts of northeast Utah, Capitol Reef, Arches, and Zion National Parks.
Common Fossils: Belemnites, a variety of dinosaur species

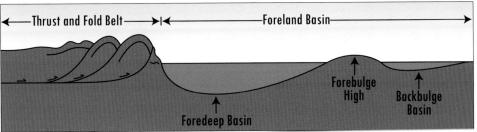

Figure 4.1 *Compressional tectonic forces created an elevated mountain belt containing thrust faults and folds. The weight of the growing orogeny weighs down the partly buoyant crust in adjacent areas creating a foreland basin which contains the foredeep, forebulge, and backbulge. Thrust belts typically do not involve Precambrian crystalline basement rocks and are referred to as "thin-skinned" tectonic deformation. No scale inferred. For comparison see Fig. 6.1, pg 47.*

Examples

▶ Early Jurassic rocks of Utah are world renowned and include the Wingate, Kayenta, and Navajo Sandstones of the Glen Canyon Group. Many of the salmon and white colored cliffs of Utah's Colorado Plateau are comprised of these formations. The Wingate and Navajo Sandstones were primarily deposited by wind-blown sand dunes that covered vast areas of Utah and surrounding states (Fig. 4.2). The Kayenta Formation that is enveloped by the Wingate and Navajo, represents river channels and associated deltaic deposits. River systems of the Kayenta reworked and redeposited the underlying eolian sand deposits of the Wingate Sandstone before eolian dunes again spread over the landscape as the Navajo Sandstone. The Navajo is picturesque and is one of the most highly studied formations on Earth. Relatively rare zircon crystals found within the quartz-rich Navajo can be used to interpret the **provenance** of these Jurassic sandstones. Studies suggest that the rocks from which the sand grains were derived were as far away as the Appalachian Mountains. The individual sand grains were moved from the Appalachians by river systems running northwest into southern Canada and were then blown southward into Utah. The Navajo blanketed much of Utah. Its maximum thickness can be found near Zion National Park where it is thought to be approximately 670 meters thick. In southern Utah the Navajo forms the "White Cliffs" of the Grand Staircase (Fig. 4.3). The Navajo Sandstone has abundant connected pore spaces between the sand grains which gives it excellent properties for becoming a subsurface reservoir for oil and natural gas. For example, the Navajo Sandstone serves as the **reservoir rock** for oil in the Covenant field in the central Utah thrust belt.

Middle Jurassic rocks in Utah include the eolian Temple Cap Formation, the Carmel Formation, the Entrada Sandstone, and their time equivalent rocks (Fig. 4.5). The Carmel Formation represents a shallow marine incursion across much of Utah while the Entrada Sandstone is again comprised of eolian sandstones and arid alluvial and **sabkha** deposits. From east to west, the Entrada transitions from eolian dunes that helped form the arches in Arches National Park (Fig. 4.6), to sandstones and mudstones that form the **hoodoos** at Goblin Valley State Park, to slightly recessive red mudstones and sandstones in central Utah that were formed by alluvial plains (see Paleogeographic Reconstruction of the Entrada Sandstone, pg. 30). As sandstone (a reservoir rock) transitions into mudstone (a non-reservoir, **sealing rock**), fluids moving through the subsurface such as oil and natural gas may become trapped. The pinch-out of reservoir rocks into non-reservoir rocks is called a stratigraphic **trap**. Folds and faults may create structural traps. Sometimes stratigraphy and structure combine to form combination traps. There is ample evidence that Utah's Entrada Sandstone has trapped hydrocarbons in numerous structural and combination traps (Fig. 4.7).

The most famous Late Jurassic rocks of Utah are the dinosaur-bearing strata of the Morrison Formation. The Morrison was deposited by rivers, floodplains, and lakes that ran from the western Cordillera of Nevada into western Colorado and beyond (Fig. 4.4). Dinosaurs survived on these alluvial plains that were at times both lush with vegetation and occasionally dry with drought. Drought likely created the spectacular dinosaur bone accumulations at the famous Cleveland-Lloyd Quarry in the San Rafael Swell wherein the dinosaurs became mired in mud as they sought water in a dwindling floodplain pond. Bones at Dinosaur National Monument were likely swept up and buried by rivers that had returned to the area after severe drought conditions had decimated local dinosaur populations. Climatically, the Morrison Formation serves as a transitional phase from the dry eolian conditions of the Early and Middle Jurassic to more temperate and humid conditions of the Cretaceous Period.

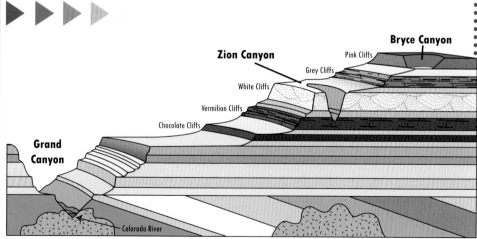

Figure 4.3 *The Grand Staircase: When a succession of stratified sedimentary rocks is tilted and exposed at Earth's surface it differentially erodes into a series of cliffs and slopes, like steps on a stairway. The Grand Staircase was eroded by the Colorado River and its many tributaries. Two of Utah's national parks are hosted within the Grand Staircase including Zion National Park, formed in the White Cliffs, and Bryce Canyon National Park, formed at the top of the staircase in the Pink Cliffs.*

Navajo Sandstone

Figure 4.2 *The Navajo Sandstone represents a sea of wind-blown sand called an erg. Dunes more than 100 meters high covered Utah during this time. Upon burial, the Navajo Sandstone has become a reservoir for both water and petroleum.*

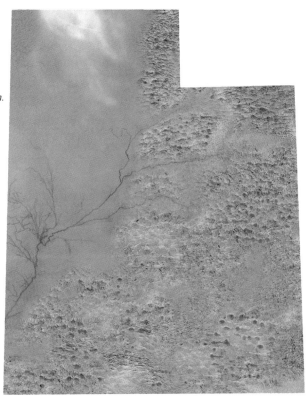

Morrison Formation

Figure 4.4 *The climate during the dinosaur-bearing Morrison Formation was at times relatively humid and arid, like the present-day Okavanga Delta, Botswana, Africa*

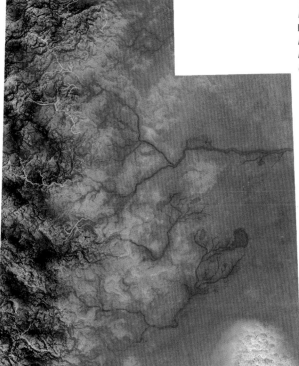

Did you know... dinosaur species found in the Morrison Formation have been used as references for species created in the film industry such as *Jurassic Park* and *Jurassic World,* although many dinosaurs in the movies are actually from the Cretaceous Period.

Stratigraphic Column - *Capitol Reef National Park*

JURASSIC	Morrison Formation	Brushy Basin Mbr	River floodplain	
		Salt Wash Ss Mbr	River channels	
		Tidwell Mbr	Alluvial plain	
	San Rafael Group	Summerville Formation	Coastal mudflat	
		Curtis Formation	Shallow marine	
		Entrada Sandstone	Eolian to sabkha	
		Carmel Formation	Shallow marine to coastal	
		Page Ss	Eolian dune	
	Glen Canyon Group	Navajo Sandstone	Eolian dune	
		Kayenta Formation	Braided streams	
		Wingate Sandstone	Eolian dune	

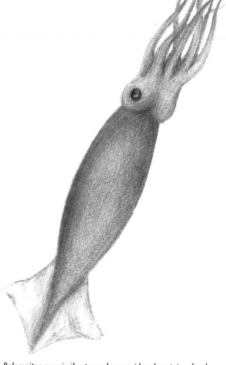

Belemnites are similar to modern squid and contain a hard internal skeleton, ink sac, hard beaks, and tentacles that point forward. The most commonly preserved part of these powerful swimmers is the guard. The guard is a tapered cylindrical structure that provided a counterweight to the buoyancy made by a chambered shell.

Figure 4.5 Southwest portion of Zion National Park. View is across the Virgin River (low foreground) and covers red Triassic rocks of the Moenkopi at the base through the Middle Jurassic Temple Cap Formation on the skyline. The massive pink-white cliffs are the Navajo Sandstone and represent the White Cliffs of the Grand Staircase. Photo courtesy of Ken Hamblin.

Jurassic World Map

★= Utah

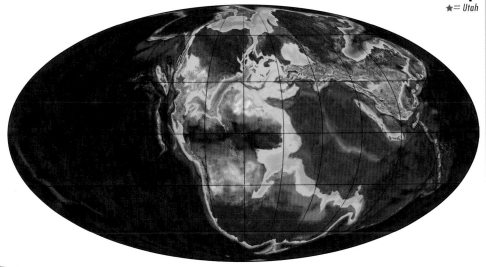

Did you know...

dinosaur bones and

petrified wood are often

radioactive and contain

uranium-bearing minerals.

The *Allosaurus* was the primary predator in North America during the Jurassic. Most known specimens of this carnivore appear in the Morrison Formation. On average they grew to about 8.5 meters long and were capable of killing healthy large herbivores.

Figure 4.6 Northward view of the Devil's Garden area of Arches National Park. The salt-cored Salt Valley Anticline (left quarter) helped to fracture the wind-blown sandstones of the Entrada Sandstone (red center) and Moab Member of the Curtis Formation (white right). The fracturing and subsequent erosion created the rock fins which were the precursors to the thousands of arches in the park. Photo courtesy of Ken Hamblin.

Figure 4.7 *Thin section of the Middle Jurassic Entrada Sandstone, southeast San Rafael Swell. The darker blue represents pore space between mainly quartz sand grains. The black represents tar or residual oil indicating that this formation likely trapped oil and/or gas in the subsurface before it was exposed to the atmosphere by uplift and erosion.*

0.5mm

Late Cretaceous
100-66 MA

Orogeny & Interior Seaway

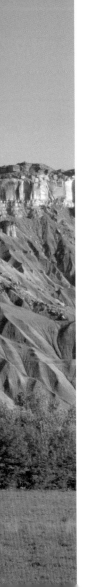

▶ The North American continental plate continued moving westward and converging with the eastward moving portion of the Pacific oceanic crust called the Farallon Plate. As this convergence rate increased during the Jurassic, the stage was set for a major orogeny. This mountain-building event culminated around 100-66 Ma, toward the end of the Cretaceous Period (145-66 Ma). Due to east-west compression, a north-south trending mountain belt developed from Mexico through Canada. In Utah, this mountain building event was called the **Sevier Orogeny,** named after the area of the Sevier River in central Utah. From a relatively thick succession of sedimentary rocks in western Utah, a series of low angle **reverse faults** (thrusts) shoved layers of strata eastward, often placing older rocks on top of younger rocks. Geologists estimate that some strata were pushed more than 50 kilometers to the east. As the layered rocks slid along the detachment surface of the thrust fault, the overlying layers were susceptible to anticlinal folding when the angle of the thrust ramped upwards or terminated. These **anticlinal** folds can trap petroleum in economic accumulations such as the Covenent oil field near Sigurd, Utah. Compressive forces also exist in front of the thrusted rocks creating a fold belt comprised of anticlines and **synclines**. Together the thrust and fold belt culminated in an elevated mountain system that is estimated to have risen more than 3 kilometers above sea level. This orogeny was yoked to a sedimentary basin to the east. The **Western Cretaceous Interior Seaway** inundated eastern Utah. This seaway eventually connected the Gulf of Mexico to the Arctic Ocean along the eastern edge of this vast thrust and fold belt.

Gray Cliffs of the Mancos Shale near Capitol Reef National Park. Gray shale represents the relatively deep rocks of the Western Cretaceous Interior Seaway whereas the beige sandstones represent the western shoreline. See World Map on pg. 43.

Late Cretaceous - Paleogeographic Reconstruction

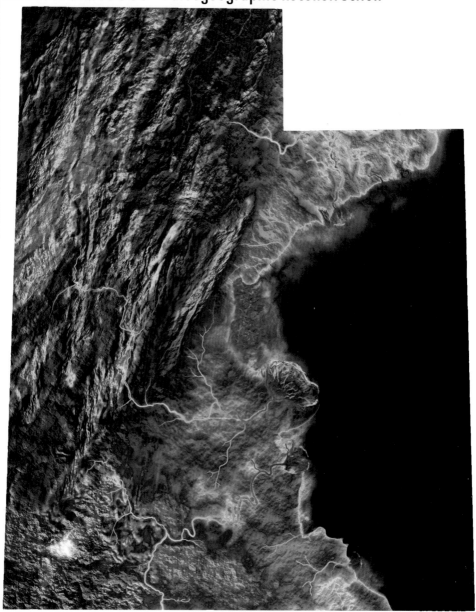

Mancos Shale (~70-80 Ma)

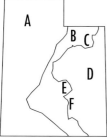

A Sevier Orogenic belt
B Coastal plain
C Vernal Delta
D Western Cretaceous Interior Seaway
E Last Chance Delta
F Notom Delta

Fast Facts

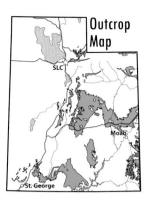

Outcrop Map

Age: Cretaceous Period 145-66 Ma
Climate: Warm temperate
Key localities: Capitol Reef and
Bryce Canyon National Park areas, Wasatch Plateau/
Castle Cliffs, Book Cliffs
Common Fossils: Dinosaurs, ammonoids, snails, plants

Examples

▶ In central Utah the thrust portion of the Sevier Orogeny is evidenced by at least five major thrust faults. Successive thrusts broke out eastward from preexisting older thrusts located to the west. The last movement on the eastern faults were as young as 50-40 Ma. One easily observable thrust fault is located on the east face of the Canyon Range near Scipio, Utah, along Interstate 15 (Fig. 5.1). There Precambrian and lower Paleozoic rocks overlie middle Paleozoic rocks, thus older rocks are stacked above younger rocks. The color contrast vividly displays the location of the Canyon Range Thrust.

East-central Utah was located between the mountains to the west and the marine seaway to the east. Alluvial plains and shoreline systems preserve evidence of dinosaurs and other prehistoric creatures that thrived in coastal swamps and lagoons. These swamps and lagoons eventually became buried and created economic coal and **coalbed methane** accumulations that supply the populous Wasatch Front with energy. Superbly exposed ancient delta and wave-dominated **shoreface** systems in east-central Utah have been studied by geologists from around the world. Numerous studies have focused on the deeper marine rocks of the Mancos Shale and its deltaic member called the Ferron Sandstone. The Blackhawk Formation and the Castlegate Sandstone represent shoreface and alluvial environments, respectively.

Figure 5.1 *Westward view of the Canyon Range in central Utah illustrates the Canyon Range Thrust wherein older rocks are thrust above younger rocks. In this photo, Precambrian and lower Paleozoic rocks (brown) were thrust eastward and over middle Paleozoic Devonian rocks (light gray). Photo courtesy of Ken Hamblin.*

The components of these ancient systems are heavily studied because they dramatically preserve architectural and spatial changes to shoreline systems as sea level fluctuated through time (Fig. 5.2). A primary forcing mechanism for sea level rise and fall are cyclic changes in Earth's orbital position relative to the Sun. These orbital variations affect the **insolation** that throws Earth into warmer and colder climates. As Earth moves from warmer to cooler climates ocean water can be stored on the continents as vast ice sheets, alpine glaciers, and lakes. Simple thermal contraction of sea water can also affect sea level. Consequently, when global sea level falls shoreline systems move basinward across the continental shelves. Sea level was also affected by regional tectonics associated with the Sevier Orogeny. Geologists studying these shoreline systems use these forcing mechanisms to make qualitative and quantitative models to better predict where petroleum may have accumulated in similar systems worldwide.

Figure 5.2 Northwestward view of the Book Cliffs in east-central Utah illustrate successive marine shorelines building basinward through the Late Cretaceous. Photo courtesy of Ken Hamblin.

With a total of 15 horns and spikes on its head and frill, the *Kosmoceratops* had one of the most elaborate faces of any dinosaur. Paleontologists believe the ornate facial decorations were used to attract mates. The animal from nose to tail measured about 4.5 meters and weighed over 1000 kilograms. The skull alone measured about 2 meters.

Stratigraphic Column - *Capitol Reef National Park*

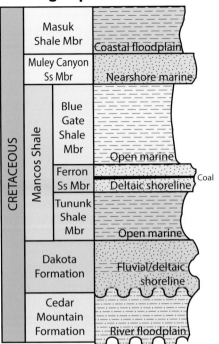

CRETACEOUS				
		Masuk Shale Mbr		Coastal floodplain
		Muley Canyon Ss Mbr		Nearshore marine
	Mancos Shale	Blue Gate Shale Mbr		Open marine
		Ferron Ss Mbr	Coal	Deltaic shoreline
		Tununk Shale Mbr		Open marine
	Dakota Formation			Fluvial/deltaic shoreline
	Cedar Mountain Formation			River floodplain

Palms were abundant in the swampy coastal plains adjacent to the Cretaceous seaway, eventually creating thick deposits of coal.

Late Cretaceous World Map

★= *Utah*

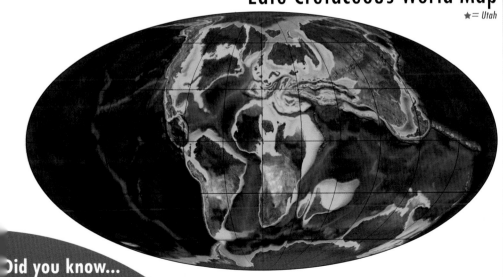

Did you know...

The shorelines associated with the Western Cretaceous Interior Seaway developed valuable energy resources including coal and natural gas.

Early Paleogene

66-44 MA

Mountain Uplifts & Intermontane Basins & Lakes

▶ Convergence of the North American continental plate and a portion of the Pacific Ocean crust, called the Farallon Plate, continued during the Paleogene (66-23 Ma), but the angle of the **subducting** oceanic plate apparently flattened. The flatter angle of the subducting plate allowed heat associated with subduction to penetrate further eastward than during the Late Cretaceous Sevier Orogeny of central Utah. This heat softened the previously brittle rocks in the lower part of the continental crust creating a different style of rock deformation relative to the Sevier Orogeny. Instead of low angle thrust faults and folds (Fig. 4.1) pushing sedimentary rocks eastward ("thin-skinned" compression), deeply buried Precambrian igneous and metamorphic rocks "popped up" vertically along high angle reverse faults ("thick-skinned" compression; Fig. 6.1). These crystalline basement-cored uplifts were first mapped in the area of Laramie, Wyoming, thus this mountain building event is called the **Laramide Orogeny**. Uplifted basement rocks caused the overlying sedimentary pile to fold. At the surface these folds express themselves as broad, usually asymmetrical anticlines and monoclines. Uplifts activated or reactivated during the Paleogene include from north to south: the Uinta Mountains, San Rafael Swell, Uncompahgre Uplift, Waterpocket Fold, Monument Upwarp and associated Comb Ridge, and the Kaiparowits Uplift. The isolated uplifts created topographic basins adjacent to and between one another. Subtropical to temperate climatic conditions provided ample rainfall creating large lakes in these basins. These lakes include ancient Lake Flagstaff (60 Ma) and Lake Uinta/Lake Green River (55-44 Ma). Rocks formed from sediments of Lake Flagstaff created the multihued hoodoos and ridges of Bryce Canyon National Park (Fig. 6.2 and facing page). Rocks from Lake Uinta have been an economic boon for northeastern Utah because they preserved organic-rich materials which serve as the source rock for numerous oil and gas accumulations. Organisms within these rocks include abundant freshwater fishes, turtles, mammals, snakes, frogs, and crocodiles.

Photo of Paleogene hoodoos in Bryce Canyon National Park. Photo taken from Bryce Point: view is to the north.

Early Paleogene - Paleogeographic Reconstruction

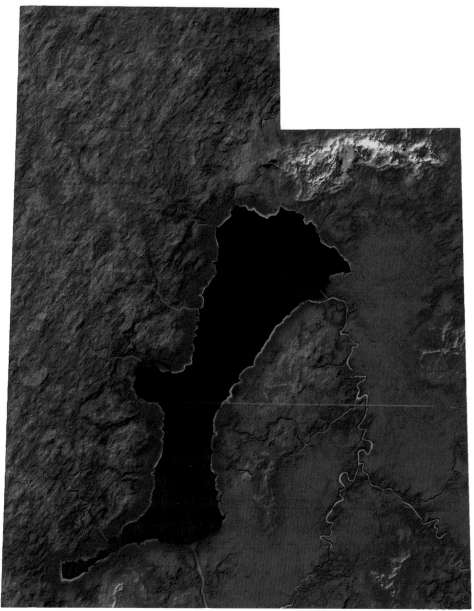

Flagstaff Limestone (~60-40 Ma)

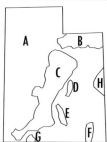

A Eroding Sevier Orogeny
B Uinta Mountains
C Lake Flagstaff
D San Rafael Swell
E Waterpocket Fold
F Monument Upwarp
G Kaiparowits Uplift
H Uncompahgre Uplift

Fast Facts

Age: Early Paleogene 66-44 Ma
Climate: Temperate
Key localities: Bryce Canyon National Park, Uinta Basin, Roan Cliffs, Northeast and Central Utah
Common Fossils: Fish, turtles

Outcrop Map

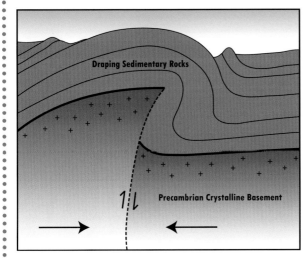

Draping Sedimentary Rocks

Precambrian Crystalline Basement

Figure 6.1 *Laramide uplifts typically involve vertical to high angle reverse fault movement of the Precambrian crystalline basement (usually metamorphic or intrusive igneous) due to compressional tectonic stresses. The structural style is called "thick-skinned" tectonics. Examples of Laramide uplifts in Utah include the Uinta Mountains, San Rafael Swell, and Waterpocket Fold.*

Examples

▶ The elongate Lake Flagstaff (60 Ma) was geographically constrained by the remnants of the Sevier Orogeny to the west, the Waterpocket Fold and San Rafael Swell to the east, and the Uinta Mountains to the north. Lake Flagstaff sediments created the Flagstaff Limestone in the north, and the orange to purple to white hues of the Claron Formation in Bryce Canyon National Park. These colors represent a variety of depositional conditions. Red to orange mudstones reflect muddy waters and shoreline deposits, purplish hues reflect exposure surfaces that created **paleosols**, and white rocks indicate deposition of calcium carbonate in clearer water that eventually created limestones and dolomites. After deposition and lithification, the Claron was fractured by several different tectonic forces. These fracture systems intersect one another at nearly 90 degree angles. As the rock was uplifted from the subsurface and exposed to the erosive agents at the surface, the fractures served as avenues of accelerated weathering and erosion creating fractured ridges and pinnacles. Also, the layered strata were comprised of different rock types which eroded differentially: some faster and some slower. Thus, the fracture systems combine with differential erosion of the strata to produce the ghostly hoodoos at Bryce Canyon.

Lake Uinta (55-44 Ma) was a subsidiary lake to a larger system of lakes that flanked the Uinta Mountains on the south, east, and north. In Utah the sediments that accumulated in and around Lake Uinta formed the Green River Formation.

Today, Green River Formation rocks are buried in the subsurface of the present day topographic basin known as the Uinta Basin. The Green River Formation has a maximum thickness of nearly 2 kilometers. These rocks also created thick accumulations of oil shale. Oil shale contains **kerogen**, an immature organic compound that needs to be processed (heated) in order to produce crude oil. Although presently considered uneconomical, the oil shale formed in these intermontane lakes is a huge resource for hydrocarbons if it was ever to become economically viable. As younger alluvial sediment was shed from the Uinta Mountains, the Uinta and Duchesne River Formations were formed. The total thickness of Paleogene rocks in the Uinta Basin eventually accumulated to well over 3 kilometers. As some of the organic-rich rocks of the Green River Formation were buried to depths exceeding 1.5 kilometers, they entered the **oil window**. The associated heat and pressure at depth started to break down the long hydrocarbon molecules of the kerogen to produce natural gas and waxy crude oil which is produced in a variety of fields in northeast Utah.

Entelodonts are sometimes referred to as 'hell pigs' or 'terminator pigs.' The pig-like carnivores stood about 2 meters tall at the shoulder and had proportionally small brains. The bony lumps on the sides of their heads are a distinguishing feature and are thought to attach powerful jaw muscles.

Figure 6.2 *The Pink Cliffs of the Claron Formation represent the upper step of the Grand Staircase. See figure 4.3 for context. They are spectaularly exposed in the Bryce Canyon Amphitheater. The Pink Cliffs form the caprock to the Colorado River drainage basin. The rim of the cliffs is eroding at approximately 0.3 to 1.2 meters per 100 years which is equivalent to 3.2 to 11.3 kilometers per million years. Photo courtesy of Ken Hamblin.*

Stratigraphic Column - *Bryce Canyon National Park*

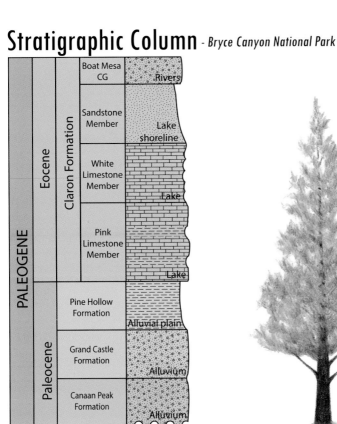

			Boat Mesa CG	Rivers
PALEOGENE	Eocene	Claron Formation	Sandstone Member	Lake shoreline
			White Limestone Member	Lake
			Pink Limestone Member	Lake
	Paleocene		Pine Hollow Formation	Alluvial plain
			Grand Castle Formation	Alluvium
			Canaan Peak Formation	Alluvium

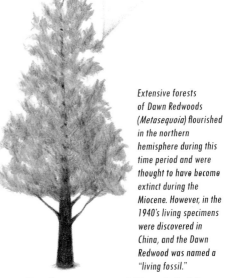

Extensive forests of Dawn Redwoods (Metasequoia) flourished in the northern hemisphere during this time period and were thought to have become extinct during the Miocene. However, in the 1940's living specimens were discovered in China, and the Dawn Redwood was named a "living fossil."

Paleogene World Map

★ = *Utah*

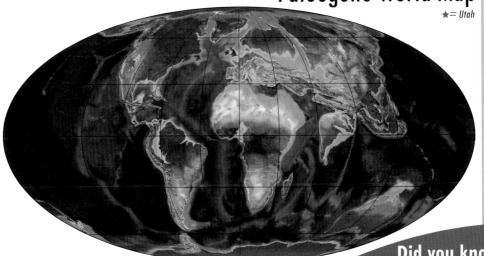

Did you know...

organic-rich sediments (kerogen) produced in Paleogene lakes surrounding the Uinta Mountains constitutes one of the largest oil shale resources in the world.

▶ ▶ ▶ ▶ ▶

49

Late Paleogene & Neogene
44–2.6 MA

Volcanism & the Basin and Range

Starting about 36 Ma, during the latter part of the Paleogene Period, the rate of convergence between the North American continental plate and the Farallon oceanic plate slowed. Eventually this caused the subducting plate to detach from the overlying continental plate. It foundered, sank, and started to roll back (**slab roll-back**). Roll-back of the subducting oceanic plate created mafic **magma** that rose into the deeper crust causing it to melt. This melting created large pools of magma in the subsurface. On occasion, these magma chambers violently erupted in western Utah and Nevada, extruding their magma to the surface to form super volcanoes (defined as having at least 1,000 cubic kilometers of **pyroclastic** material extruded in one explosive event). This magmatic material, composed of both ash and silica-rich lava flows (rhyolites), created thick successions of volcanic rocks in much of western Utah and Nevada. After expelling their materials, these magma chambers collapsed creating **calderas** up to 40 by 60 square kilometers in area. One of these super volcanos erupted as much as 5,900 cubic kilometers of pyroclastic material. This period of violent volcanic eruptions in western Utah was called the "**ignimbrite** flareup." In northern and central Utah granite-dominated stocks of various sizes are at the core of several present-day mountain ranges, some of which developed into mining districts (Fig. 7.1). In eastern Utah magma rose to the near surface and ponded in isolated mushroom-shaped bodies under a relatively thin veneer of overlying sedimentary rocks. These laccoliths are composed of granite and diorite. Laccoliths are a characteristic igneous feature of the modern Colorado Plateau.

*Crystal Peak is located in southwestern Utah. It is composed of crystal-rich rhyolitic ash (tuff) that erupted approximately 35 Ma. This event filled a paleovalley. Crystal Peak is a visual testament to the huge volume of magmatic material that was extruded in Utah's portion of the Basin and Range. Photo courtesy of Ken Hamblin.

Late Paleogene & Neogene - Paleogeographic Reconstruction

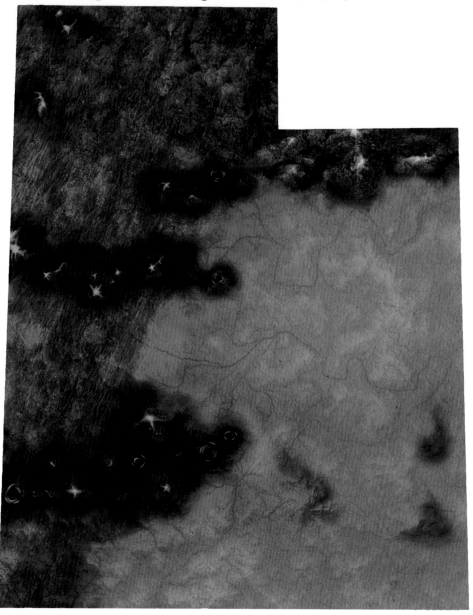

~25 Ma

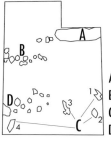

A Uinta Mountains
B Cortez Bingham volcanic belt
C Laccoliths: 1-La Sal, 2-Abajo, 3-Henry, 4-Pine Valley
D Volcanoes and calderas

Fast Facts

Age: Late Paleogene 44-23 Ma; Neogene 23-2.6 Ma
Climate: Arid high plateau
Key localities: Henry Mountains, Abajo Mountains, and La Sal Mountains (Arches National Park), Bingham Canyon Copper Mine, Marysvale, Indian Peak, southwest Utah

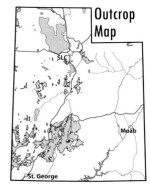

Outcrop Map

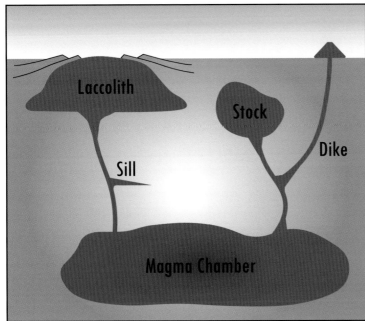

Laccolith

Stock

Dike

Sill

Magma Chamber

Figure 7.1 *A variety of igneous bodies can form from a given magma chamber. The slower the magma cools in the subsurface, the larger the crystals grow. Magma emplaced in the subsurface cools slowly, whereas magma extruded at the surface cools quickly. Thus, geologists can interpret intrusive versus extrusive igneous rocks by the crystal size.*

The laccolith is the characteristic intrusive igneous body of the Colorado Plateau. Examples include the La Sal, Abajo, and Henry Mountains. Intrusive stocks in central and western Utah often produce precious metals. Dikes and sills can be found sporadically in Utah.

The Tylocephalonyx *is a mammal related to modern horses, rhinos, and tapirs. They used their claws to grab branches and pull them down to eat. Paleontologists believe they head-butted each other with their dome-shaped skulls to show dominance.*

Examples

▶ Approximately 17 Ma and well into the Neogene Period (23 Ma – 2.6 Ma), the San Andreas fault system of California became a prominent feature along the western edge of the North American continent. This fault system resulted from the westward moving North American continental plate beginning to completely override the northeastward moving Farallon Plate and interacting with the spreading ridge of the Pacific oceanic crust. This new plate configuration, combined with continued rise in elevation due to slab roll-back, stretched the western North American crust by 20-50 percent. This extension created the Basin and Range (Great Basin) province of western Utah, Nevada, and adjacent areas to the north and south (Fig. 1.5). Extension was fast enough that the crust broke along normal faults. Eventually, bedrock units were broken and separated vertically and laterally to create topographic basins (Fig. 7.2). One wildcat oil well drilled in Utah Valley demonstrated that vertical displacement of bedrock exceeded 4 kilometers. Thus, from Utah's Wasatch Front to the Sierra Nevada Mountains of eastern California, western Utah and Nevada were extended by the development of **grabens** and **horsts**. Many of the valleys are considered to be half grabens wherein one side of the valley dropped along its normal fault considerably more than the other side. Some of the extension has been focused along previous thrust faults that created zones of weakness in rocks of the upper crust. Thus, some thrust faults formed during compression of the Sevier Orogeny became foci of relaxation and extension during the Neogene.

Characteristic volcanic rocks extruded during the past 20 Ma included basalt cinder cones and rhyolite domes associated with lava flows. This bimodal volcanic activity appears to result from decompressional melting of upper mantle rocks which in turn is associated with crustal extension. Ultimately it is the plate configuration that created these characteristic features and rocks.

Paleogene World Map

★ = Utah

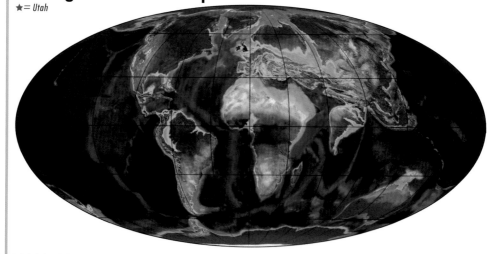

Figure 7.2 *Tule Valley is located in west-central Utah. It displays classical Basin and Range topography. The mountains are uplifted horsts while the valleys are sediment-filled grabens. Normal faults separate the horsts and grabens. View to the south. Photo courtesy of Ken Hamblin.*

Stratigraphic Column - Wah Wah Mountains

Did you know...

when super volcanoes erupt, the ash ejected from the vent can rise into the atmosphere and circumvent the Earth, often initiating short-term global climate change.

NEOGENE	**Miocene**	Blawn Formation	Mostly igneous extrusive rocks
PALEOGENE	**Oligocene**	Isom Formation	
		Three Creeks Tuff	
		Lund Formation (Needles Range Group)	
		Wah Wah Springs Formation (Needles Range Group)	
		Cottonwood Wash Tuff (Needles Range Group)	
		Escalante Desert Formation (Needles Range Group)	

Figure 7.3 *Bingham Canyon Mine is a stock containing metal-bearing ore that has been in operation since 1906. It has produced 19 million tons of copper, and accounts for 25% of US production today. The pit is one kilometer deep and 4 kilometers wide. In addition to copper, the mine produces significant amounts of gold, silver, and molybdenum. Photo courtesy of Ken Hamblin.*

In southwestern Utah and eastern Nevada, super volcanoes of the Indian Peak caldera complex erupted approximately 10,000 cubic kilometers of ash that spread over an area of 55,000 square kilometers. Compare that volume to the one cubic kilometer of volcanic material erupted by Washington State's Mount St. Helens in 1980, and one may get a feeling for the size of these eruptions! Individual eruptions have been mapped as separate formations based on their petrologic make up, including different mineral percentages and textural characteristics. Mapped formations of the Needles Range Group include the Escalante Desert, Cottonwood Wash Tuff, Wah Wah Springs, and Lund Formations. In central and northern Utah intrusive rocks dominated by granites produced numerous mining districts including the Marysvale, Alta, and Tintic districts as well as the famous Bingham Canyon open-pit copper mine near Salt Lake City (Fig. 7.3). Copper, gold, silver, and lead are often carried by hot, hydrothermal waters associated with intrusions. Some intrusions are barren of these precious metals. On the Colorado Plateau of southeastern Utah, laccoliths emplaced granitic rocks amongst sedimentary rocks (Fig. 7.1). Because granites are more resistant to erosion than most sedimentary rocks, these laccolithic intrusions have differentially eroded to create isolated mountain ranges. These igneous-cored ranges include the La Sal, Abajo, Henry, and Pine Valley Mountains and Navajo Mountain (see Paleogeographic Map inset, pg. 52).

The Wasatch Front separates the Basin and Range from the more eastward Middle Rocky Mountains and Colorado Plateau provinces (see Fig 1.5). Numerous north-south trending normal faults in the western portion of the Colorado Plateau

attest to the ongoing eastward extension of the Basin and Range. In the Basin and Range, alluvial sediments were eroded from the elevated ranges and filled the adjacent basins. Intermixed with the alluvium are a variety of basalt, ash, and rhyolite lava flows. The Salt Lake Formation in northern Utah is comprised of alluvial and lake sediments and volcanic rocks. Combined, these rocks are estimated to be more than 4 kilometers thick under the Great Salt Lake. Young volcanic flows still cover the surface of the landscape in many areas of the Basin and Range including the Black Rock Desert of west-central Utah and Pine Valley Mountains near St. George, Utah. Furthermore, geothermal springs are numerous in western Utah and Nevada. Thus, normal faults, alluvium, bimodal volcanism, and geothermal areas all attest to the ongoing extension of the Basin and Range province.

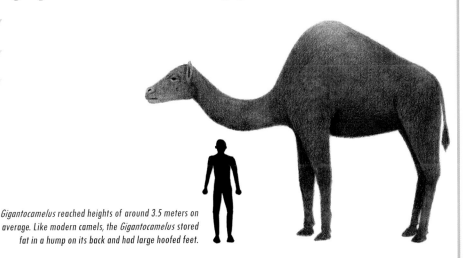

Gigantocamelus reached heights of around 3.5 meters on average. Like modern camels, the Gigantocamelus stored fat in a hump on its back and had large hoofed feet.

Quaternary
2.6 MA-PRESENT

Glaciation & Lake Bonneville

▶ The Quaternary Period (2.6 Ma–Present) includes the Great Ice Age which continues to exist today. With continents positioned in high latitudes and robust oceanic circulation, Earth was poised to accumulate large volumes of ice on its high latitude land masses. Milankovitch orbital perturbations drove climate cycles between glacial and interglacial phases (Fig. 8.1). Earth entered cooler climatic intervals as CO_2 decreased in the atmosphere. Snow accumulated during the winter seasons in high latitudes but was unable to completely melt during cool summers. Thus, snow accumulated to greater depths year after year until its own weight turned it to dense ice. Continental-scale ice sheets expanded over high latitude land masses like Canada, Northern Europe, Greenland, and Antarctica. Past glacial maxima can be discerned from the landforms they left behind and that are still present today. Evidence for Lake Bonneville, a giant ice age lake, includes wave-cut terraces and wave-built beach deposits along the flanks of present day mountain ranges.

Geoscientists have estimated that more than 3% of Earth's ocean volume was used to create these continental glaciers. Thus, during glacial growth, global sea level dropped up to 120 meters and shorelines shifted hundreds of kilometers from their highstand positions high on the continental shelves to their lowstand positions at the **shelf-slope break**. Furthermore, there is ample evidence that there have been dozens of glacial/interglacial cycles during the past 2.6 Ma. We are presently near the end of an interglacial (warm) phase of this cyclic climate system.

Photo of a mud flat within the Great Salt Lake. The fenceposts act as a wick for the salty water. As the water evaporates at the top of the post, salt crystals precipitate causing the posts to swell and splinter. Photo courtesy of Ken Hamblin.

Quaternary - Paleogeographic Reconstruction

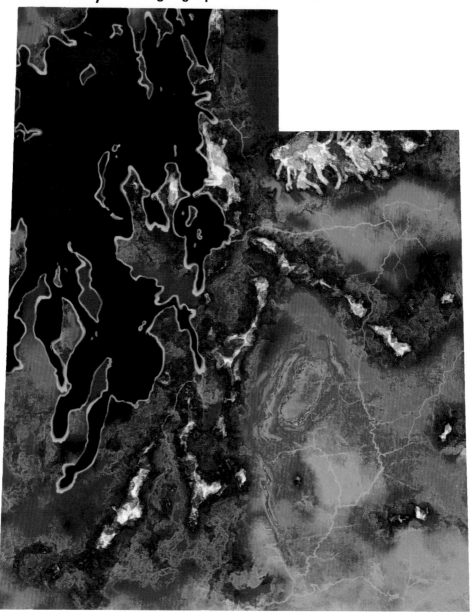

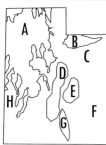

~15 thousand years ago

A Lake Bonneville
B Uinta Mountains
C Uinta Basin
D Wasatch Plateau
E San Rafael Swell
F Colorado Plateau
G Waterpocket Fold
H Basin and Range

Fast Facts

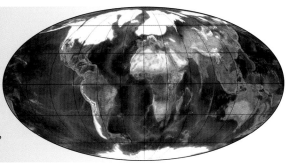

Age: Quaternary 2.6 Ma-present
Climate: Temperate to arid
Key localities: Antelope Island State Park, Great Salt Lake, Bonneville Salt Flats
Common Fossils: Brine shrimp, ground sloths, mammoths

The driving mechanisms for the Ice Age were elucidated by a Serbian mathematician named Milutin Milankovitch (1879-1958). Milankovitch, working from previous studies by Scotsman James Croll (1821-1890), made astronomical observations that led him to conclude that there were three major changes in the shape and angle of Earth's orbit around the Sun (Fig. 8.1). According to the Milankovitch theory these three variations in Earth's orbit affect insolation and, combined, can throw Earth in and out of ice ages by cyclically changing Earth's heat budget. Ice ages, in turn, can cause dramatic changes in global sea level.

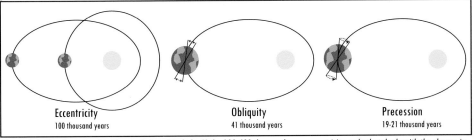

Eccentricity	Obliquity	Precession
100 thousand years	41 thousand years	19-21 thousand years

Figure 8.1 *Milankovitch orbital "perturbations" include: 1) the 100-400 thousand year eccentricity cycle that deals with the change in the eccentric (elongate) orbit of Earth around the Sun; 2) the 41 thousand year obliquity cycle that deals with the change in angle (tilt) of the Earth's spinning axis relative to the orbital plain of the Earth around the Sun; and 3) the 19-21 thousand year precession cycle dealing with the "wobble" of Earth's spinning axis.*

The Megalonyx is a medium-sized ground sloth that lived in North America during the Pleistocene. They could reach heights of up to 3 meters. Their tail and flat-footed hind feet enabled them to stand somewhat upright to reach into trees for food.

Did you know...
when Earth enters a cold glacial maximum condition, approximately 3% of Earth's ocean volume is stored on the continents as giant ice sheets.

Examples

▶ During the Pleistocene, large Ice Age lakes (pluvial lakes) covered many of the basins in Utah and Nevada. These lakes resulted from precipitation and temperature changes associated with continental ice sheet growth during glacial maximums. A relatively recent pluvial lake, Lake Bonneville, formed approximately 26 thousand years ago (ka). Lake Bonneville's level fluctuated due to both climate change and cataclysmic events. One such event occurred when a natural dam at Red Rock Pass, Idaho, cataclysmically failed allowing Lake Bonneville to partially drain into the Snake River. A famous geologist, G. K. Gilbert (1843-1918), first documented Lake Bonneville and its fluctuations by recording the numerous wave-cut and wave-built terraces and other geomorphic features along the edges of mountain ranges. At the time, many ranges were islands rising above this great freshwater lake. As Lake Bonneville dried up, largely from climate change associated with Milankovitch cycles, salt was concentrated in isolated bays. Eventually, the Bonneville Salt Flats were created. The modern Great Salt Lake is a remnant of the much larger Lake Bonneville.

Did you know...

tiny brine shrimp (~1 cm long) are one of the only species to inhabit the salty water of the Great Salt Lake. Brine shrimp eggs (or cysts) are harvested for sale as fish and prawn food.

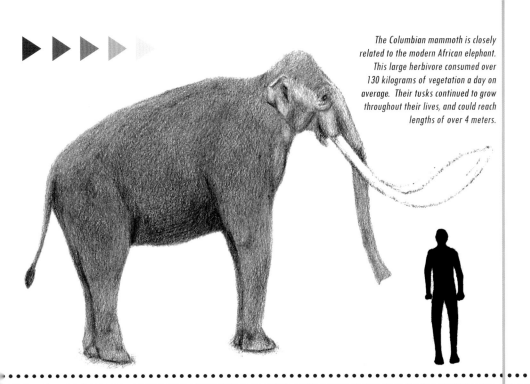

The Columbian mammoth is closely related to the modern African elephant. This large herbivore consumed over 130 kilograms of vegetation a day on average. Their tusks continued to grow throughout their lives, and could reach lengths of over 4 meters.

The Great Salt Lake, a remnant of Lake Bonneville, loses water only through evaporation. Because of this and its shallowness, the Great Salt Lake's areal extent varies dramatically from a historic low in 1963 of 4400 km² to a historic high in 1988 of 8500 km². This variation also affects the salinity of the lake which fluctuates from 50-250 parts per thousand. Normal ocean water is 35 parts per thousand. Photo courtesy of Zac Durrant.

Summary

Precambrian-Lower Paleozoic (2500 – 323 Ma)

- *Tectonic Setting*: Development of the North American continent through accretion (2500 Ma); failed-rift basin develops in Utah as ancient supercontinent Rodina splits (750 Ma); continental shelf sedimentation dominates along the passive margin of North America's western edge (541-323 Ma).
- *Climatic Setting*: Largely unknown

Pennsylvanian & Permian (323-252 Ma)

- *Tectonic Setting*: Convergence of supercontinent Gondwana and northern hemisphere continents to form Pangea causes compression which creates uplifts and yoked basins (e.g., Uncompahgre Uplift and Paradox Basin). Northwestern Utah is on continental shelf of North America.
- *Climatic Setting*: Arid conditions in Utah dominate but sea level fluctuates due to glacial/interglacial cycles of continental ice sheets in southern Pangea.

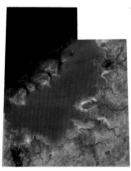

Triassic (252-201 Ma)

- *Tectonic Setting*: Closure of landmasses into supercontinent Pangea is completed and orogenesis ends. Oceans become toxic and create the Permian-Triassic mass extinction of marine species. Utah displays gentle relief as it passively moves northward.
- *Climatic Setting*: Northward movement of Utah from the equator into the trade-winds belt (10-30 degrees north latitude) causes the climate to change from humid to arid during the Triassic.

Jurassic (201-145 Ma)

- *Tectonic Setting*: Utah continues to move northward and also westward relative to the Pacific Ocean Plate. East-west convergence causes island arcs systems to accrete to western North America and creates the Cordilleran Arc. A foreland basin begins to develop east of the Cordillera.
- *Climatic Setting*: Arid climates dominate due to Utah residing in the trade-winds belt through much of the Jurassic.

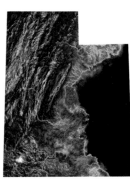

Late Cretaceous (100-66 Ma)

- *Tectonic Setting*: North America's westward convergence with the Farallon Plate of the greater Pacific Ocean crust creates greater compressive forces which culminate in the Sevier Orogeny, a fold and thrust belt (thin-skinned tectonic style). Thrust sheets load the crust regionally and create an eastward foreland basin that fills with marine water to create the Western Cretaceous Interior Seaway.
- *Climatic Setting*: Warm, temperate climates should have existed as Utah has moved north of the trade-winds belt but high altitudes make it relatively dry.

Early Paleogene (66-44 Ma)

- *Tectonic Setting*: The angle of the subducting Farallon Plate flattens. The Laramide Orogeny, with characteristic Precambrian basement involvement (thick-skinned tectonic style), dominates eastern Utah.
- *Climatic Setting*: Temperate climates produce enough rainfall to fill intermontane basins forming large lake systems.

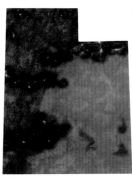

Late Paleogene & Neogene (44-2.6 Ma)

- *Tectonic Setting*: The Farallon Plate's convergence rate slows which allows it to sink. Slab roll-back initiates the "ignimbrite flareup" that produces super volcanoes and giant calderas in western Utah. Eastern Utah experiences the emplacement of laccolithic intrusions. At ~17 Ma the North American Plate begins to override the Pacific mid-ocean ridge initiating the San Andreas Fault system of California and Basin and Range extension.
- *Climatic Setting*: Dry highland with intermontane lakes.

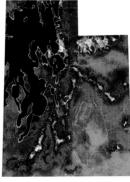

Quaternary (2.6-present)

- *Tectonic Setting*: Basin and Range extension continues and expands eastward into the western edge of the Colorado Plateau.
- *Climate Setting*: Temperate to arid. Cooler phases of the climate cycle produce alpine glaciers and large pluvial lakes (i.e., Lake Bonneville). Earth is presently at the end of an interglacial (warm) phase of the Ice Age.

Alluvial

Pertaining to sediment deposited by a river, stream, or flowing water.

Anoxic

Refers to water that has little or no dissolved oxygen.

Anticline

A crustal fold, usually induced by tectonic forces, in which the limbs dip away and downward from the fold axis (fold hinge or center-line). Anticlinally folded sedimentary rocks that have been eroded flat display successively older rocks toward the axis of the fold. Anticlines are capable of trapping large accumulations of petroleum in the subsurface.

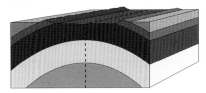

Arkose

A coarse-grained sandstone rich in feldspar and quartz that usually has been deposited relatively close to its source.

Braided River

A network of several small, branching, and interlacing streams in a wide, shallow, sand-rich channel belt, usually due to an overload of sediment and seasonally variable flow.

Caldera

A large circular depression formed by the collapse of a volcanic magma chamber after an eruption.

Carbonate Rock

A rock composed mainly of carbonate minerals such as calcite ($CaCO_3$) and dolomite ($Ca,MgCO_3$).

Coalbed Methane (CBM)

A form of natural gas that is extracted from coal seams. CBM gas can be used as an energy resource.

Continental Shelf

The gently sloping portion of the continental margin that is between the shoreline and the continental slope. See *Shelf-slope Break.*

Convergence

Occurs when two tectonic plates, commonly one continental and one oceanic, move towards each other. See *Subduction.*

Delta(ic)

A body of sediment deposited at the mouth of a river (e.g., Mississippi River delta). The sediment accumulates at a delta because the river has descended to base level and loses its energy and thus its ability to transport sediment.

Diapir

Plugs of ductile rock such as salt or gypsum that passively deforms into vertical columns in the subsurface due to sediment loading. Salt movement, in turn, deforms the adjacent strata.

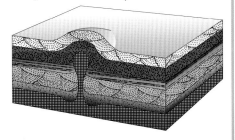

Downcut

Stream erosion cuts down vertically into the strata (as opposed to laterally).

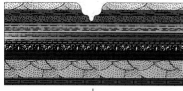

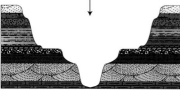

Entrenched River

A river that has incised into the underlying bedrock, usually because of local uplift or from a significant drop in base level.

Eolian

Refers to processes and areas (usually desert environments) where wind is the primary mode of sediment transport and deposition.

Ephemeral

A stream or lake that is short-lived or that only flows briefly in response to local precipitation. Also a short-lived event.

Era

A chronologic unit made up of periods. Examples include the Paleozoic Era (ancient life), Mesozoic Era (middle life), and Cenozoic Era (modern life).

Erg

Refers to a specific type of eolian depositional environment. A vast area covered in sand dunes (e.g., Africa's Sahara Desert). It is often described as being a "sea of sand."

Erosion

The removal and transport of sediment, after it has been weathered from the rock. Water is the most prolific agent of erosion; however, other agents of erosion include wind, ice, and gravity.

Facies

A rock unit that has characteristics that distinguish it from adjacent units and reflect the origin (depositional environment) of the rock.

Gondwana

A continent in the southern hemsiphere during the late Paleozoic that resulted from the breakup of the supercontinent Pangea.

Graben

An elongate block of bedrock that has been lowered relative to the surrounding rocks by normal faults creating a topographic basin. The adjacent block that sits higher than the graben is referred to as a "horst." See *Horst*.

Highstand

Refers to the period of time when sea level has flooded the continental shelf between cycles of sea-level change. See *Shelf-slope Break*.

Hoodoo

Rock columns, chimneys, or pinnacles possessing bulging knobs and narrow recesses. They are common in Bryce Canyon National Park and were formed by differential erosion along fractures of the strata.

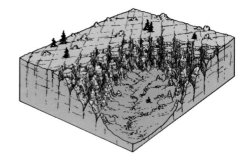

Horst

An elongate block of bedrock that has been raised relative to the surrounding rocks by normal faults, creating a topographic high/mountain range. The adjacent block that sits lower than the horst is referred to as a "graben." See *Graben*.

Hydrologic cycle

The continuous movement of water, on, above, and below the surface of Earth.

Ichnofossil

A fossilized structure such as tracks, trails, burrows, borings, etc. that preserves the life activities of a creature, but does not include the creature itself. Also called a trace fossil.

Ignimbrite

Another name for hot ash-rich material that flowed sometimes violently from a volcanic vent and was then consolidated and welded to rock as it cooled (an ash-flow tuff).

Insolation

Incoming solar radiation hitting Earth from the Sun. Changing insolation can drive the ice ages forming intervals of cold glacial maximums and relatively warm interglacials.

Kerogen

A fossilized organic material found in sedimentary rocks, especially shales, which can be heated and processed to form petroleum products.

Laccolith

A mushroom-shaped mass of igneous rock that never made it to the surface of Earth. It is flat-floored and has caused the overlying strata to be uplifted in a dome shape.

Laramide Orogeny

A mountain building event during the Late Cretaceous and early Paleogene period (~75-45 Ma) that created portions of the Colorado Plateau and Rocky Mountains. The event was caused by compressional plate tectonic forces wherein oceanic plates west of North America slid into and under the North American continental plate. Laramide orogenesis involves Precambrian basement rocks: considered a "thick-skinned" tectonic style. See Fig. 6.1.

Laurasia

A supercontinent in the northern hemisphere during the Mesozoic that resulted from the breakup of the supercontinent Pangea.

Lowstand

Refers to the period of time when sea level has retracted from the continental shelf between cycles of sea-level change. See *Shelf-slope Break*.

Magma

Molten rock material generated within Earth that can rise to the surface or near surface through cracks in the crust. When solidified, magma forms igneous rocks.

Mass Extinction

Periods in Earth's history when abnormally large numbers of species die out within a short time.

Monocline

An asymmetrical fold, usually tectonically-induced, in Earth's crust in which one flank of the fold is steeply inclined and the other is only gently tilted or flat. See Fig. 2.3.

Normal fault

A fault in which the hanging wall has moved down relative to the footwall. Associated with extension.

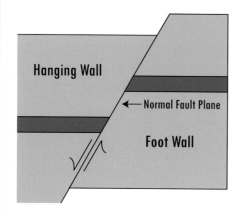

Hanging Wall

← Normal Fault Plane

Foot Wall

Oil window

Organic matter matures and oil generates at specific depths and temperatures in Earth. This special environment is referred to as the oil window.

Oolitic

A rock composed of ooids, which are small sand-sized spherical grains made of concentric layers of calcium carbonate.

Orogeny

A mountain building event.

Paleosol

An ancient soil horizon produced by a subaerial exposure surface.

Pangea

A supercontinent that existed from about 300 to 200 Ma. Our current continents are a result of fragmentation of Pangea. It was separated into two smaller supercontinents called Laurasia and Gondwana that were in the northern and southern hemispheres, respectively.

Panthalassic Ocean

The ocean that surrounded most of Pangea during the late Paleozoic and early Mesozoic.

Passive Margin

A continental boundary without plate-boundary tectonism.

Peat

Loose sediment made by partially carbonized plant remains that were decomposed in a water-saturated environment.

Plate Tectonics

Massive forces caused by the movements of Earth's crustal plates. These forces deform rocks by inducing fractures, faults, and folds. It is these same forces that separate continents, create mountains and volcanoes, and continue to shape Earth's surface.

Potash Salts

Refers to potassium chloride (KCl). Broadly, it is a term for water-soluble potassium salts. Today it is mined and commonly used as a fertilizer.

Provenance

The place of origin or the location from which the sediment in a sedimentary rock was derived.

Pyroclastic

Clastic rock material that was formed during the eruption of a volcano.

Reservoir Rock

A term used in petroleum geology that refers to porous and permeable rock that yields oil or gas. Common types include sandstone, limestone, and dolomite.

Reverse Fault (Thrust)

A fault in which the hanging wall rides above the footwall. Reverse faults with a low angle (<25°) fault plane along much of its lateral extent is called a thrust fault.

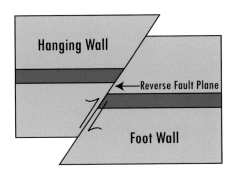

Sabkha

A sedimentary environment formed in arid to semiarid conditions on coastal plains just above the normal high-tide level. Sabkhas typically contain evaporite minerals, tidal-flood, and eolian deposits.

Salt Dome

The upward movement of a salt diapir that can produce a three-dimensional, dome-shaped structure in the overlying sedimentary rocks. See *Diapir*.

Salt Walls

Elongate accumulations of salt in the subsurface wherein salt flow is focused through fractures or faults due to lateral loading (like squeezing a tube of toothpaste). In southeastern Utah, some of these subsurface features are up to 3 kilometers high, 5 kilometers wide, and 110 kilometers long. Arches National Park results, in part, because of the emplacement of salt walls.

Sealing Rock

A term used in petroleum geology that refers to the impermeable rock that traps hydrocarbons and prevents them from migrating upward.

Sevier Orogeny

A mountain building event during the Cretaceous that involved folding and eastward thrusting of rocks along what is now the eastern edge of the Great Basin in Utah. Sevier orogenesis does not involve Precambrian basement rocks: considered a "thin-skinned" tectonic style. See Fig. 4.1.

Shelf-slope Break

The point wherein the continental shelf increases in dip toward the ocean's abyssal plain.

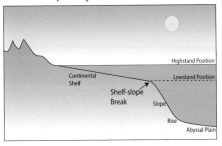

Shoreface

The sloped area basinward from a lake or marine shoreline. It is a relatively high energy environment due to waves.

Siliciclastic

Refers to clastic non-carbonate rocks which bear large amounts of silicon minerals such as quartz.

Slab Roll-back

A process wherein the subducting plate (usually oceanic) begins to increase its subduction angle as it decends into the mantle.

Subduction

The process of one tectonic plate descending beneath another, usually due to differences in mass.

Superimposed

Refers to a drainage system that downcuts through the formations on which it has developed onto a different structure that lies unconformably beneath.

Syncline

A fold, usually induced by plate

tectonic forces, in which the limbs dip toward the axis (hinge or center-line) of the fold. In folded sedimentary rocks that have been eroded flat, the younger rocks are toward the fold axis or hinge.

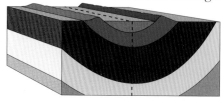

Tectonism

See *Plate Tectonics*.

Thrust Fault

See *Reverse Fault*.

Trade-winds Belt

A system of warm, usually dry winds that are associated with high pressure cells that move from the subtropics (~23-38° N and S) to the equator.

Trap

A barrier that prohibits the movement of oil or gas and allows accumulation. Includes reservoir rock and seal rock.

Unconformity

A time gap in the geologic record that marks where a rock unit is overlain by another unit that is not next in chronologic succession. An unconformity is produced by either non-deposition or erosion.

Uplift

Plate tectonic forces often cause rocks to be elevated relative to their previous position and adjacent land surfaces. The uplifted rocks may then experience increased rates of erosion. Tectonic uplift has produced the Colorado Plateau.

Volcanic Arc

A generally curved linear belt of volcanoes produced above a subduction zone.

Weathering

The process of physically breaking, or chemically altering rocks by exposure at Earth's surface. This is not to be confused with erosion (although they work together), as erosion involves the removal and transportation of the weathered material.

Western Cretaceous Interior Seaway

A large inland sea during the Late Cretaceous and into the early Paleogene that separated North America into two landmasses. The sea stretched from the modern day Gulf of Mexico up through the United States and Canada to the Arctic Ocean.

References

Blakey, R. and Ranney, W., 2008, Ancient Landscapes of the Colorado Plateau, Grand Canyon Association, 156 p.

Hintze, L.F. and Kowallis, B.J., 2009, Geologic History of Utah, Brigham Young University Geology Studies Special Publication 9, Scott M. Ritter, ed., 225 p.

Morris, T.H., Ritter, S.M., and Laycock, D.P., 2012, Geology Unfolded: An Illustrated Guide to the Geology of Utah's National Parks, Geology Unfolded, LLC, 72 p.

Sprinkel, D.A., Chidsey, T.C., Jr., and Anderson, P.B., eds., 2010, Geology of Utah's Parks and Monuments, Utah Geological Association Publication 28, 3rd Edition, Utah Geological Association and Bryce Canyon Natural History Association, 623 p.

See other GEOLOGY UNFOLDED, LLC publications:

- *GEOLOGY UNFOLDED: An Illustrated Guide to the Geology of Utah's National Parks*, T.H. Morris, S.M. Ritter, D.P Laycock, 2012

- *Exploring the Diverse Geology of Capitol Reef National Park*, T.H. Morris and K.G. Spiel, 2016

Notes:

Notes:

About the Authors

Thomas Morris

Tom received his B.S. in geology from Brigham Young University (BYU) and his M.S. and Ph.D. from the University of Wisconsin-Madison. He spent four years in the oil and gas industry working in Louisiana. He made his way back to BYU as a clastic sedimentologist and stratigrapher in 1990. Tom has spent more than 25 years studying and teaching about the sedimentary rocks of Utah. He has published more than 35 articles on the rocks of the Colorado Plateau. His passion is showing both undergraduate and graduate students the wonders of sedimentary geology ... and Utah is one of the best places on Earth to do that!

Kinsey Spiel

Kinsey was attending Brigham Young University-Idaho when she took a geology course to fulfill a general education science requirement. She soon fell in love with geology and changed her major to Earth Science Education. She received her B.S. and then began pursuing a M.S. in geology at BYU. Still interested in education, she started working with Dr. Tom Morris to write and design books aimed at helping the public better understand geology, specifically in Utah's National Parks. A native of the Pacific Northwest, Kinsey grew up exploring and still loves hiking and enjoying the outdoors.

Preston Cook

Although Preston is a second generation geologist, it wasn't until his undergraduate degree at BYU that he came to embrace his father's science and reluctantly admit that the apple doesn't fall far from the tree. As an undergraduate, Preston found he was most passionate about sedimentology and stratigraphy and channeled his passion into pioneering new techniques for creating paleogeographic maps, which are on display in this book. He has recently completed a M.S. at BYU. He is pursuing a career in the oil industry. As a graduate student, he has received grants from the Utah Geological Society as well as the Society for Sedimentary Geology.

Hannah Bonner

After a childhood spent living at Boy Scout Camps and many summers working in Utah's red rock deserts, Hannah is hooked on exploring the natural world. Consequently, as an undergraduate student at BYU, majoring in geologic sciences is the perfect fit. Hannah has a passion for understanding the deep history of her surroundings. Further, she takes great satisfaction in interpreting geologic academia in a manner relevant to a broad audience. Illustrating paleogeographic maps of Utah for this book has perfectly combined her love for geology, education, and graphic design.

Acknowledgements

We wish to express our sincere appreciation to the geoscientists and discoverers who have contributed to the cumulative knowledge of Utah's geologic history. Without their efforts this book could not have been written. It is humbling to think of the man hours of field and laboratory work that has been completed over the past century to get us to our present level of understanding. We salute all those souls who dedicated parts of their lives in this effort.

This book is the result of our attempt to synthesize major events of Utah's geologic past for the benefit of students and those who are geologically curious. We have synthesized these short chapters from numerous resources including books by Blakey and Ranney (2008), Hintze and Kowallis (2009), Sprinkel et al., eds., (2010), and our own book Morris et al. (2012). We recommend these books to readers who wish to delve deeper. Plate continental reconstructions were purchased from Ron Blakey's Mollewide global maps, Colorado Plateau Geosystems, Arizona. Prehistoric life figures were drawn by Elizabeth Lew. Geology students George Jennings, Shelby Johnston, Scott Meek, and Jeff Valenza assisted in developing several early drafts of the paleogeographic reconstructions. Hannah Bonner was instrumental in developing the final detailed versions of the paleogeographic maps. Her talents and efforts are greatly appreciated. We appreciate input and review from colleagues Eric Christiansen, Scott Ritter, and our good friend Tom Chidsey from the Utah Geological Survey. Photographs were taken by the authors except where noted. This book also includes knowledge derived from untold papers on the sedimentology, stratigraphy, volcanic, and igneous activities of Utah's geologic past. The book also incorporates ideas from faculty and student colleagues and our own research experience.

Finally, we thank our families and friends who supported us in this fun endeavor.